SUPER COOL SCIENTISTS #2

A Story and Coloring Book Celebrating Today's Women in Science

Written by Sara MacSorley
Illustrated by Yvonne Page

Keep being super cool!
Sara MacSorley

DEDICATION

Super Cool Scientists is dedicated to all the women out there, right now, doing amazing work in science, technology, engineering, and mathematics. It is for all the women reaching back to help others along their paths. We see you. We appreciate you. We thank you.

Super Cool Scientists is dedicated to all the young people out there who are curious about how the world works, who like to take things apart and put them back together, who can't stop exploring their backyards. It is for all the young people who think science is exciting and that they can use it to change the world - you can!

Super Cool Scientists is also for my people. The people who supported me through my journey - starting with and circling back to science. My mom, who supported my interests from an early age, who validates my occasional choice to eat ice cream for dinner, and is still the first person I call when I have a problem. My dad, who checks in on me in his own unique way which usually involves poop emojis. My teachers who fostered my curiosity. My mentors who opened my eyes to what I could do with a science degree beyond research. It is for all the women in science in my life who pushed me and continue to inspire me with their intelligence, compassion, and wisdom. Thank you.

Super Cool Scientists Community
Share your #supercoolscientists coloring artwork or stories on Twitter @SuperCoolSci
or on Facebook on the Super Cool Scientists Page.

A Super Cool Scientists Production
www.supercoolscientists.com

ISBN-13: 978-1723483905
ISBN-10: 1723483907

INTRODUCTION

Super Cool Scientists #2 is a story and coloring book that celebrates the incredible stories of twenty-two dynamic women working in science, technology, engineering, and mathematics (known as STEM). STEM is super cool. It's around us every day - in the technology we use, the processes that grow the food we eat, and the natural phenomenon that literally keep us grounded on the planet.

STEM is where it's at - projects that aim to solve problems, solutions that help people and our environment, opportunities to work with people all over the world, and usually, good paying careers. However, women and people of color are still considerably underrepresented in STEM.

Today's science is complex and requires the different expertise of many people in many places. Projects need to be approached from a multidisciplinary point of view and creative solutions come from including a range of perspectives. We are limiting our opportunities for scientific advancement if we are leaving anyone out. It's time to make some changes and include everyone at the table.

There are things we can all do to help, even if we aren't scientists. We can make contributions as teachers, parents, journalists, consumers, and employers. Our small way to support this effort of inclusion in STEM is to celebrate women in science to inspire the next generation of researchers, educators, and communicators.

The women highlighted in this book are astronomers, microbiologists, cybersecurity engineers, and veterinary researchers. They travel the world, explore unknown environments, and let ocean sediments even take them back in time. Their work highlights that science careers are broad and exist far beyond the white lab coat.

These women are entrepreneurs, artists, educators, authors, and mothers. They are leaders, mentors, advocates, and friends. Their lives are made up of many pieces, just like ours.

These women represent different ages, experiences, races, origins, abilities, orientations, and religions. They are proof that science is for everyone.

We want to empower women currently in the STEM space. We want to recognize the incredible work they do and share it with others. We want to tell them they are doing a great job and to keep up the good work of supporting others along the way. We want them to know they are role models and inspirations. We see you doing so super cool, keep it up!

We want to inspire the next generation of young people to explore STEM careers as options for themselves. We want them to know that, regardless of potential challenges, science can be a career for you. We are here and will be your community. We will support you.

SCIENTISTS

Alicia Morgan
Bachelor's in Aerospace, Aeronautical, and Astronautical Engineering from Tuskegee University
Master's in Industrial Engineering from New Mexico State University

Amanda Preske, Ph.D.
Bachelor's in Chemistry from the Rochester Institute of Technology
Doctorate in Chemistry from the University of Rochester

Anne Estes, Ph.D.
Bachelor's in Zoology and Wildlife, Marine Biology from Auburn University
Master's in Biological Sciences from Auburn University
Doctorate in Ecology and Evolutionary Biology from the University of Arizona

Camille Eddy
Bachelor's in Mechanical Engineering from the University of Idaho (in process)

Cherisa Friedlander
Bachelor's in Marine Biology from the University of Rhode Island
Master's in Applied Marine and Watershed Science from California State University - Monterey Bay

Emille Davie Lawrence, Ph.D.
Bachelor's in Mathematics from Spelman College
Doctorate in Mathematics from the University of Georgia

Jesse Shanahan
Bachelor's in Arabic Linguistics and Philosophy from the University of Virginia
Master's in Astronomy and Astrophysics from Wesleyan University

Kashfia Rahman
High School (in process)

Lucianne Walkowicz, Ph.D.
Bachelor's in Physics from Johns Hopkins University
Master's in Astronomy from the University of Washington
Doctorate in Astronomy from the University of Washington

Margaret Adriatico
Bachelor's in Biology and Chemistry from Christopher Newport University

Michele Yatchmeneff, Ph.D.
Bachelor's in Civil Engineering from the University of Alaska - Anchorage
Master's in Engineering Management from the University of Alaska - Anchorage
Doctorate in Engineering Education from Purdue University

Mónica Feliú-Mójer, Ph.D.
Bachelor's in Biology from the University of Puerto Rico - Bayamón
Doctorate in Neurobiology from Harvard University

Nicole Acevedo, Ph.D.
Bachelor's in Animal and Poultry Science from Virginia Polytechnic Institute and State University
Master's in Reproductive Physiology from Virginia Polytechnic Institute and State University
Doctorate in Molecular and Integrative Physiology from the University of Michigan

Pamela McCauley, Ph.D.
Bachelor's in Industrial Engineering from the University of Oklahoma
Master's in Industrial Engineering from the University of Oklahoma
Doctorate in Industrial Engineering from the University of Oklahoma

Parisa Tabriz
Bachelor's in Computer Science from the University of Illinois at Urbana-Champaign
Master's in Computer Science from the University of Illinois at Urbana-Champaign

Prasha Sarwate Dutra
Bachelor's in Chemical Engineering from the University of Pune in India
Master's in Mechanical Engineering from the University of Texas - Arlington

Sarah Myhre, Ph.D.
Bachelor's in Biology from Western Washington University
Doctorate in Climate Change, Oceanography, and Climate Communication from the University of California - Davis

Sunshine Menezes, Ph.D.
Bachelor's in Zoology from Michigan State University
Doctorate in Biological Oceanography from the University of Rhode Island

Sylvia Acevedo
Bachelor's in Engineering from New Mexico State University
Master's in Engineering from Stanford University

Tiffany Lyle, Ph.D., DVM
Bachelor's in Biological Sciences from the University of Georgia
Master's in Anatomic Pathology from Purdue University
Doctorate in Veterinary Medicine from the University of Georgia
Doctorate in Comparative Biomedical Sciences from Purdue University

Tracy Chou
Bachelor's in Electrical Engineering from Stanford University
Master's in Computer Science from Stanford University

Trisha Walsh
Bachelor's in Sociology and Social Work from the Massachusetts College of Liberal Arts
Master's in Computer Information Systems and Quantitative Business Methods from California State University - East Bay

ALICIA MORGAN

Alicia Morgan's career is a great example of the value of STEAM - bringing in the arts into science, technology, engineering, and mathematics. It is all about creative problem solving.

Alicia is an engineer by training. She studied Aerospace Engineering as an undergraduate at Tuskegee University and then Industrial Engineering in graduate school at New Mexico State University. She worked at large engineering companies including Lockheed Martin, The Boeing Company, and Raytheon. At Boeing, she helped track engine failures. Did you know the Boeing 777 airplane has three types of engines?! Alicia's job was to come up with improvement plans for issues that came up during testing. She had to use her creativity to find solutions.

Aerospace Engineering is a super cool field not just for the engineers, but for everyone to learn about since it is such a part of our history. Air and spacecraft have literally transformed how we travel around the world and out of this world.

As a kid growing up in New Orleans, Alicia enjoyed dissecting different processes to learn more about how things worked and how they might be improved. She went to science programs after school and those experiences furthered her interest in engineering. Her parents always encouraged her learning and passed on their love of reading and writing. Alicia also enjoyed public speaking. She was the speaker at her high school graduation and now speaks at conferences around the country. Alicia uses those stages to share the message that scientists are equally analytical and creative.

After years of working in the corporate sphere, she expanded her outreach activities to working with education non-profits. Now, she spends her time showing people the value of STEAM and loves that she found a way to use her analytical and creative skills. Alicia designs activities about aviation and spaceflight for schools and the public at the Frontiers of Flight Museum. She gets to work with programs that are helping students discover their talents and then help those students along their own paths. Alicia loves the arts too - creative writing, poetry, live music, film, and more. Working at a museum is the perfect place for combining passions for science and art.

Alicia remembers being more of an introvert early in her career - hard to believe knowing she speaks in front of such large audiences! It took her time to learn how to speak up for herself and to build relationships with mentors. Having a diverse community of mentors in your corner can be really helpful in working through professional and personal challenges. Part of life, regardless of your field, is failure. How we handle those situations is what defines us. Alicia's work with students encourages the growth mindset - the idea that we can all continue to move forward and learn from any situation. We're never totally stuck and we can find creative ways to take flight and reach our goals.

It's all about creative thinking to solve problems. Scientists are creative and artists are great problem solvers. We're not so different and Alicia helps us remember our common ground.

AMANDA PRESKE

Many scientists, including Amanda Preske, grow up making things. Amanda grew up in the suburbs of Syracuse, New York, making jewelry and lots of crafts as a kid. Her dad had a workshop and her mom had a sewing room so there was always something new she could explore creating.

Amanda grew up in a science household. Her mom, and eventually her brother too, studied biology and her father was an engineer. She remembers some of her middle school science teachers making a big impact on her desire to go further into science. You could say they were catalysts. Amanda went on to study chemistry because of the field's hands-on nature. She studied at the Rochester Institute of Technology and got her doctorate degree at the University of Rochester. She took a lot of art classes in college too.

Her graduate work focused on the creation of quantum dots - tiny crystals that have the ability to emit energy. The size of the quantum dots determines their properties such as their color. The study of quantum dots is important because they can be used to make super efficient solar panels and help scientists cure disease. Her research process in the lab was reminiscent of her creative process outside of the lab.

Amanda knows that creativity fuels new ideas, so it is critical in art and in science. She was inspired by her high school art teachers who let her explore new mediums and then started a science-themed jewelry company, Circuit Breaker Labs, while she was in college. The jewelry is created using broken circuit boards and resin - a super cool way to repurpose old computer parts. Her jewelry always sparks a reaction. Amanda uses that to talk about electronics waste and recycling with her audience. Only about 30% of electronics waste is recycled, so we definitely have room for improvement.

Now, the company is her full-time career and she enjoys sharing her love of science through wearable art. She has to be careful in the jewelry creation as circuit boards are made of some nasty stuff like fiberglass and heavy metals (so don't try this at home!). Once everything is cleaned and covered in resin, the jewelry is totally safe to wear and it is gorgeous. Pretty cool considering she started making jewelry when she was nine years old. As a business owner, Amanda is in charge of everything from product research and design to sales and marketing. All those responsibilities can be tough but at the same time empowering because she is in charge of the company's success.

Her chemistry degrees come in handy as Amanda does a lot of science outreach activities. She volunteers at the local science museum to show kids how nanomaterials, liquid crystals, and thin films work. She collaborates with other scientists and creators on local events like a Science Expo and an urban craft market. Plus she loves to travel! She has visited Morocco, Ghana, and South Africa.

Amanda wants others to see that creativity is important in art, science, and in business!

ANNE ESTES

Anne Estes is a biologist who studies the microbes that live inside and on dung beetles - yes, poop beetles.

She specifically looks at microbe-host interactions because they can affect the nutrition, health, behavior, and even the genetic makeup of each partner. A host can be any kind of organism - humans, other animals, plants, insects. Microbes are tiny and they are everywhere. Anything microscopic is a microbe like bacteria, viruses, even some very small fungi. The collection of all the types of microbes living in a certain host is called a microbiome. Humans have a microbiome of over 10,000 types of microbes - many of them in our gut.

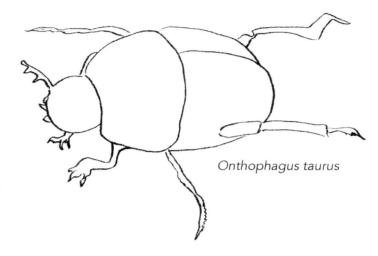

Onthophagus taurus

Anne is interested in how the diet of the dung beetles at each life stage affects the insects and the microbes. Disruptions and disturbances in host-microbe interactions in other insects can cause issues for flying, having babies, and protecting themselves from predators. Dung beetles, like many insects, eat different items and change body shapes at different stages of their lives. As larvae they chew on plants and as adults they feed on liquids. Think about how a caterpillar becomes a butterfly - there are a lot of structural changes that happen during that metamorphosis. The microbes living inside the insect are there the whole time, so they are dealing with their entire house changing around them while they're in it! Anne's research questions how the microbiome changes during these different life stages in order to survive.

Besides her research, Anne spends a lot of time teaching college students. She and her husband are both Biology Professors at Towson University. They are truly partners in work and in life. Anne also does a lot of outreach work to show others the importance of microbes in our everyday lives. You can find her working at a local citizen science lab in Baltimore, hosting online workshops for healthcare professionals, or blogging at *Mostly Microbes*. She loves finding creative ways to help people understand the usually invisible world of microbes.

Anne grew up in a small town in Alabama. Her family has actually lived in that area for over 175 years! She and her sister spent a lot of time exploring the creek near their home and helped out often in the family store, a small plumbing and electrical business. She loved helping her dad fix things and listening to him tell stories as they were driving to service calls. Anne's mom was the first person in her family to go to college, she studied English. She would teach classes at local colleges, do the bookkeeping for the family business, and nurture Anne's interest in science by exploring new streams and visiting the library to read research papers.

Sometimes life comes full circle. As a young girl, Anne participated in Girl Scouts. She loved how she got to learn about the world while also learning how to help make the world, and her community, a better place. Now, she is a troop leader for her daughter's Girl Scout group helping young girls explore their interests. Anne can see her father's super storytelling skills and her mother's determination in her two daughters. Now, she is the one telling her kids that they can, and should, do whatever they want for a career.

CAMILLE EDDY

Camille Eddy is a mechanical engineer who is super dedicated to getting more young women interested in technology.

Mechanical engineers look at all the physical parts of a product and think through how they need to fit together in order to function. Think about your cell phone for example. Cell phones have a lot of parts and components. Different engineers work on each specific issue like how the phone charges, how the phone feels in your hand, and how the phone connects to a computer to share or backup data.

An important part of mechanical engineering is product testing. They not only design the physical hardware of a product, they also create a prototype of the product and put it through extensive testing. They test to make sure all the pieces fit together, that the product does what it is supposed to do, and if it breaks, they figure out how to fix it. The tests themselves also have to be designed by engineers like Camille.

Camille loves the problem solving aspect of product design and prototyping. There is something exciting about approaching a totally new challenge and being the first person to design possible solutions. She also enjoys connecting with other technology professionals. Camille makes time each week to get together after work with other tech interns. She says, you are who your friends are. Her group of tech friends grew up in different areas and studied a variety of subjects in school and they all come together through their love of technology and giving back. Camille also spends time each week mentoring other young women in STEM and planning STEM events.

As much as Camille loves talking to other people, as a young student she was terrified of public speaking. These days, she spends a lot of time talking to others about student leadership, design thinking, and cultural bias in the artifical intelligence community. Her writing and speaking engagements have allowed her to travel the country and meet lots of interesting people - including President Obama.

When Camille was a young girl growing up in Idaho, she remembers looking up to astronauts Barbara Morgan and Mae Jemison. Seeing Mae Jemison specifically gave her the representation she needed as a black woman to see herself working in the space arena. Her mom, who had a degree in business, homeschooled Camille and her sister which allowed them to really explore and research their interests. Her mom was also a big Star Trek fan so she was helpful in encouraging Camille's new found interest in learning about space.

It was a little scary to make her first move out of Idaho but Camille knew that she needed to take risks to get what she wanted for her career. Being open to new opportunities has led to even more exciting things for Camille and that is a reminder that with drive, resilience, and surrounding yourself with supportive people, you too can be a super cool scientist.

CHERISA FRIEDLANDER

Lieutenant Cherisa Friedlander is an officer in the NOAA Commissioned Corps - one of the seven United States uniformed services like the Navy or Coast Guard. NOAA stands for the National Oceanic and Atmospheric Administration and officers like Cherisa serve as specialists at sea, on land, and in the air. Her specialities are marine and climate science.

Marine and climate science are important because everything we do can be connected back to the planet, including the ocean. Earth gives us our most basic needs for survival such as the air we breathe, the water we drink, and the food we eat. Did you know every other breath we take can be traced back to oxygen produced by plankton in the ocean? Tiny plants, super important role! Understanding how natural processes work can help us understand how to make sure all the things we love about the planet are around for the future.

Cherisa has supported fisheries research at the National Marine Fisheries Service. She goes out on huge research vessels, boats over 200 feet long, where the crew uses big nets to collect the ocean animals. The crew identifies all the creatures they collect and measures and counts everything they find. Sometimes you get surprises like a jellyfish in the middle of digesting a fish!

The job of a NOAA Corps Officer is different from day to day, week to week, and you literally get to see the world. Every few years, Cherisa gets a new assignment at a different location. She has been stationed across the country from coast to coast - from her home state of Massachusetts to California. She has sailed on the Atlantic and Pacific Oceans and through the Panama Canal. During the creation of this book, Cherisa embarked on an exciting new assignment that took her to Antarctica! She visited research laboratories in Colorado and Alaska first for training where she learned how to use equipment that measures information about the atmosphere and climate.

Cherisa comes from a service family. Her father is retired Navy and one of her four brothers is a Navy Lieutenant. The travel associated with a service career is super exciting but can also be tough. You can't always be near your loved ones. She knows the time she shares with her friends and family is very special and takes advantage of that time as much as she can. The lifestyle is worth it to her to do such interesting work in marine science.

A visitor came to Cherisa's first grade classroom to talk about marine science. They brought a bunch of seashells and baleen from a whale for the kids to see and that hooked the young Cherisa - line and sinker. She had amazing science teachers in school and got to work with some super cool marine science organizations in college like the New England Aquarium and Save The Bay. She got her Marine Biology degree from the University of Rhode Island and went on to get a graduate degree in Applied Marine and Watershed Science from California State University at Monterey Bay.

EMILLE DAVIE LAWRENCE

Emille Davie Lawrence is a geometric topologist. That means she studies the mathematics which can help us understand the shape of our world, and even the universe.

Think about it. How would we know that Earth was shaped like a sphere instead of a donut without pictures from space? The answers to questions like these - along with numbers, equations, and patterns - help topologists understand the shape of three-dimensional (or 3-D) space and more.

Emille works as a Professor of Mathematics and Statistics at the University of San Francisco. Teaching classes and working with students take up a lot of her time during the week. She also spends time each week on her research like her work in spatial graph theory which looks at how collections of points and edges can be positioned symmetrically in a 3-D space. Emille's favorite parts of her job are meeting new students and learning new mathematics.

She has always loved math. She remembers being a young girl in Georgia adding up the numbers she saw everywhere - like on digital clocks and license plates. Emille noticed patterns in the numbers and her interest to understand more about what the patterns meant led her to keep learning about math. She studied Mathematics at Spelman College and then got her doctorate degree from the University of Georgia.

In graduate school, students have an advisor who helps guide them through their research and professional path. Advisors play a big part in a student's experience. Emille's first advisor wasn't super supportive, she even suggested that Emille pick another career. However, Emille was able to find other professors who believed in her and helped her pass all three qualifying exams that are required of doctoral students in mathematics. Sometimes we all need a little reminder that we have what it takes to succeed. Finding the people who can assist and support us along the way helps to balance the equation.

One of Emille's college mentors was Dr. Sylvia Bozeman. She taught her linear algebra which is the study of solving linear equations, as well as real analysis which is the theory of real numbers and real-valued functions. Dr. Bozeman is one of the founders of the Enhancing Diversity in Graduate Education summer program and she got Emille to apply. That experience introduced Emille to many other women in mathematics who remain friends and colleagues today. She pays it forward now as a graduate student mentor and instructor in the program.

STEM is all in the family. Emille's dad was a high school science teacher. She met her husband when he was studying computer science and he is now a software engineer. Time will tell if their two young children, a girl and a boy, will also go into STEM. Being a mom keeps Emille busy too. She loves spending time with her family doing things like listening to live music with her husband and exploring San Francisco with her kids. Emille also makes sure to have time to herself. She takes fitness classes and still finds time for a ballet class once in a while to fuel her love of dance.

JESSE SHANAHAN

Jesse Shanahan is a science communicator who studies extragalactic astronomy and physics. Astronomy is the study of what exists and what happens in outer space and extragalactic means outside of the Milky Way galaxy (where Earth is located). Jesse really likes space - so much so that she named her service dog Hubble after the telescope!

Believe it or not, many of the discoveries made in astronomy help us in our day to day lives. The surprising things we learn from space lead to new technologies that help improve things here on Earth, like medicine. Astronomy is interdisciplinary too, meaning you use all sorts of science like physics, math, computer science, chemistry, and even geology.

Jesse is studying Astrophysics in graduate school, works as a lab assistant, and is also a writer. She feels that communicating science is a super important part of being a scientist. Her work can often be done from home - research, writing and editing, networking, outreach, and teaching - and that is helpful as Jesse has a rare genetic condition called Ehlers-Danlos Syndrome or EDS.

EDS inhibits mobility and causes chronic pain. Sometimes, Jesse wears braces or walks with a cane. Her service dog, Hubble, helps her with both her physical and mental disabilities (she has anxiety, depression, and attention deficit hyperactivity disorder too). Prioritizing her health is important to Jesse and this means her weeks also involve medical appointments, going to the gym, time to organize her medication, and therapy. Time management is key! And she makes sure to spend plenty of time with her family and enjoying her hobbies.

Outside of work, Jesse has a lot of interests. She is a volunteer emergency medical technician, a gardener, a video gamer, and she is currently learning about woodworking and electronics. Exploring all these interests keeps Jesse curious - a critical part of being a scientist.

Jesse had parents who encouraged her curiosity when she was young. Her dad was in the military so they moved around a lot. Her mom is an elementary school math teacher who remembers Jesse having science interests as a kid like wanting to look at all the leaves from the backyard under the microscope. However, Jesse didn't go into science right away as a career. She actually started in the humanities. Such a dramatic change in focus was challenging, especially because some people think that if you study humanities you can't do math and science. Well that is simply not true - you can do both - and Jesse is a great example of that art and science mix.

She is also proof that female and disabled people can be scientists. Early on, she faced sexism and ableism. Jesse often looked to her maternal grandfather for inspiration. He kept her grounded and made her laugh when things got tough. She has a tattoo as a daily reminder of his advice and his sense of humor. Everyone belongs in STEM if that is where they want to be - even if that isn't always the message we hear. You can shoot for the stars too!

KASHFIA RAHMAN

The brain fascinates Kashfia Rahman. As a child, she observed the behaviors of a friend who had autism and began watching television segments that talked more about neuroscience. She wants to dedicate her career to the study of nervous system development and function. Specifically, Kashfia wants to be a cognitive neuroscientist with a focus on child and adolescent behavioral health.

Kashfia explores the human brain and behavior through research projects. Her international award-winning science fair project studied risky behavior in teenagers. The amygdala is the part of the brain that controls emotions and it plays an important role in risk-taking behavior. She looked at how repeated exposure to risk decreases the perception of risk. So basically, the more times you do a risky thing, the less risky it seems.

She's also had research projects about stress in teens and the effects of cell phone separation on teen health. Researching issues specific to teens is important to Kashfia because adolescents are a large but often overlooked

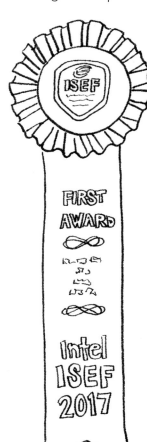

population when it comes to mental health issues. However, research shows that the teenage brain, especially the prefrontal cortex, isn't fully mature until 25 years old! The prefrontal cortex controls thinking, reasoning, impulsiveness, and more. So it is super important to consider how the teenage brain develops and influences behavior because adolescents are vulnerable. By focusing on the unique structure of the teenage brain, she hopes to raise public awareness, reduce stigma, and someday influence policy around mental health.

When she was little, Kashfia got to live in a lot of places - Indiana, Toronto, and, Kentucky. She mostly grew up in South Dakota with her family. Her dad is a Pharmacy Professor and her mom was an English teacher before she had children. Growing up in such small towns was sometimes difficult because there wasn't much diversity. Kashfia wants to help make the mental health profession more diverse so that all people are a little more comfortable seeking help for their issues. She is passionate about teaching the next generation of young scientists to continue research in neuroscience.

Kashfia has always looked up to her sister, Zarin, who is also a scientist studying neurobiology. Zarin got into science after learning more about childhood epilepsy. Another woman Kashfia looks up to is Michelle Obama. Her humanitarian work has made Kashfia want to help others. When she's not in school or working on her research projects, Kashfia volunteers at the local children's library, spends time with her close friends, and reads a lot. A part of the brain that is super important for reading is the supermarginal gyrus. It helps the brain connect letters into words!

LUCIANNE WALKOWICZ

Lucianne Walkowicz is an astronomer who studies the stars. Studying the stars can help scientists understand where life might exist beyond our own planet. She uses tools like huge telescopes and data from planetary rovers to look for clues about planetary habitability - basically if we think life could exist on other planets. She frames it as searching for "prime alien real estate." The principles of physics and chemistry are the same throughout the universe so that helps us make comparisons to what we know here on Earth.

Lucianne's research revolves around the Kepler Telescope - an amazing tool that scans 150,000 stars every half hour. She specifically looks at the light that the stars produce and can use that data to gather information on the habitability of the surrounding planets. Kepler has found over 2,300 planets so far! She is also a science leader in the Large Synoptic Survey Telescope mission. This huge telescope is being built on top of a mountain in Chile. It will house the world's largest digital camera, allowing it to capture huge images and lots of data over the ten year project. Lucianne will also work with young students to teach them how to analyze the huge amounts of data that come out of the project.

Her day job lets her do lots of different types of tasks. She does research as well as outreach. Lucianne helps create events at The Adler Planetarium to share space science with visitors. One of the favorite parts of her job is being able to share her love for understanding the universe with others and she does it well. She's active on social media, does lots of public speaking, and has even served as a star expert in a film documentary.

Lucianne grew up in New York City, with a few years in Los Angeles, California. She always liked figuring out how things worked and was interested in lots of types of science. Her first research experience happened during high school. She got her undergraduate degree at Johns Hopkins University studying physics and then got into astronomy in graduate school at the University of Washington. Astronomy is super cool because it uses many fields of science so you get more big bang for your STEM buck.

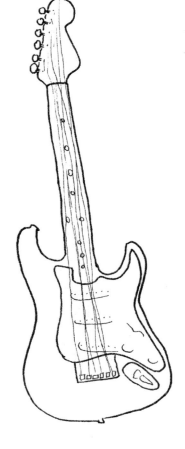

There were times in college that Lucianne wasn't sure science was for her. She had a lot of great research mentors who supported her in tough times, and a college professor who told her not to quit science. A simple comment can be enough to make you think about something differently.

Outside of work, Lucianne explores her active and artsy sides. You can often find her on her bicycle or paddleboard. Her newest activity - aerial silks! It's an artform that looks like a combination of dance and gymnastics but done in the air hanging from long pieces of fabric. She makes music, draws comics, and produces a variety show with her partner - set in space of course. Her art often uses data from her science, like using the sound of stars to make music.

Science is accessible to everyone. Stellar sharers like Lucianne come up with creative ways to make that a part of their work - on and off the job.

MARGARET ADRIATICO

What do Wonder Woman and engineers have in common? They're both heroes! Margaret Adriatico loves working with scientists and engineers to solve problems, save lives, and make the world better.

Growing up in Ohio, Margaret was interested in how and why things worked. Always armed with a screwdriver, pen-knife, and can of oil, she loved to fix things around the house like squeaky doors and loose knobs. She also started experimenting with ingredients in her family's kitchen. Cooking and baking involved a lot of chemistry, and the results of most experiments were edible and tasty. For example, the thermal degradation of polysaccharides with protein inclusions is also known as yummy peanut brittle. In other words, when you heat up certain types of sugars and add nuts, you get a liquid that hardens into the teeth-sticking candy we know and love.

In college, Margaret studied Biology and Engineering with a focus in Materials Science. Biology is the study of living things, like people, plants, and animals, and engineering is the study of structures, materials, or systems. Studying both positioned Margaret to become a super problem solver.

Margaret's dream job was to be a NASA scientist. When she graduated college, she got the chance and was selected to design new materials for the International Space Station. More than ten years after she started working at NASA, the International Space Station launched, and Margaret could see her work as reality. Her work developing materials that covered wiring on the Space Station made sure that the astronauts would be safe and could complete their missions. The material reduced the risk of fires caused by the electrical systems.

After NASA, Margaret worked as an engineer at Procter & Gamble, a company that makes everything from detergent to baby diapers. There, she developed new materials that were either biodegradable or could dissolve in water. Biodegradable and dissolvable materials reduce the amount of waste that goes into landfills. Less waste is a big deal because we produce so much trash. According to the Environmental Protection Agency, people in the United States make 251 million tons of trash a year - that is over four pounds per person, per day!

Over her engineering career, Margaret received five patents for the materials she invented. A patent is a special license from the government that allows only her to have the rights to the material design and production - it's a way to protect her inventions so she gets the credit.

Now, Margaret manages teams of scientists and engineers and she is a leader in the Society of Women Engineers, a professional organization that supports women engineers and engineering students around the world. She wants to make sure that every girl who dreams of becoming a superhero scientist or engineer can achieve her dream so we can continue making life better for everyone.

MICHELE YATCHMENEFF

Oncorhynchus nerka

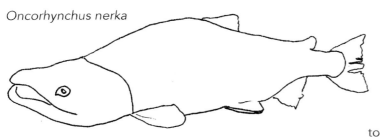

Michele Yatchmeneff is a civil engineer and advocate for helping other Alaska Natives pursue STEM degrees. Civil engineers design and build complex infrastructure like roads, buildings, and wastewater systems.

Michele knew from an early age that she wanted to go to college to get a good job but wasn't sure in what area. In middle and high school she took lots of math and science classes. One of her high school chemistry teachers suggested a summer camp in Colorado. That experience exposed Michele to engineering and set her on her STEM career path.

After studying Civil Engineering at the University of Alaska at Anchorage, she worked in the construction and engineering industry. Her engineering projects involved designing water and sewer systems in remote villages.

Now, as a Professor at the University of Alaska at Anchorage, each week at work is different for Michele. She teaches engineering courses, holds office hours to meet with students, and collects engineering education research. She makes time to meet with students in the Alaska Native Science & Engineering Program, too. The Alaska Native Science & Engineering Program works with students starting in sixth grade and supports them all the way through doctoral degrees in STEM. Michele participated in the program as a student, directed the program for several years, and has the program's creator, Dr. Herb Schroeder, as one of her mentors. Giving back to her community is important to Michele.

She grew up in King Cove and False Pass, small villages in the Alaskan Aleutian Islands. She went to school in Anchorage, the largest city in the state, about 1,000 miles away from her hometowns. Michele had to take a three hour plane ride on two different planes or take a three day ferry ride to get back to her hometowns. Over the summers, she would come home to live a traditional Unangax subsistence lifestyle and prepare her favorite foods with her family. A subsistence lifestyle is one where you collect and prepare your resources from the land in ways that respect nature and cultural traditions.

The Native people of Alaska are split into eleven cultures each with their own languages and traditions. Her parents grew up in the Aleutians and are both Unangax Alaska Native. Michele's parents work at the Aleutian Housing Authority, which oversees housing opportunities for the people who live on the Aleutian Islands. Their sense of service got passed down to Michele as did their work ethic. They also made sure that Michele and her twin brother knew the importance of education and family.

Michele says the best part of her job is working with students plus it is a job that allows her the means and flexibility to spend time with her family. Her two nephews are her favorite people in the world. She and her partner have two dogs, Sandy and Suzy. The whole family likes to get together to prepare and enjoy traditional foods like salmon fish pie and fried bread.

MÓNICA FELIÚ-MÓJER

Mónica Feliú-Mójer uses her training as a neurobiologist—a scientist who studies the biology of the brain and the nervous system—to educate and empower communities that are underserved by and underrepresented in science.

Growing up in northern rural Puerto Rico, Mónica spent her childhood catching lizards and taking care of a cow in her backyard. She loved being surrounded by nature and wanted to learn all about how things worked in her environment. She had amazing parents who continuously encouraged her curiosity.

Science and scientists seemed foreign to her though and Mónica didn't know that she too could become a scientist. None of the scientists she learned about in books or on television looked or sounded like her. Plus, as a kid she didn't have the opportunity to meet any scientists in real life.

Mónica first met a scientist in person while in college. That scientist was her biology professor, and she happened to also be Puerto Rican and a woman. Her professor challenged Mónica, fostered her curious mind, and exposed her to what being a scientist could look like for her. She was the one that got Mónica to try research in a summer program and that changed her whole career path.

She studied Human Biology at the University of Puerto Rico in Bayamón and then went on to get her doctorate in Neurobiology from Harvard University. Mónica was inspired to study neurobiology by her father. When she was eleven years old, he was diagnosed with a mental illness. Though she didn't understand his condition well, she knew it was related to the brain. This motivated her to want to do scientific research about how the brain works. His condition continues to inspire her career in interesting ways. For example, she learned that you need to listen with empathy and put yourself in the other person's shoes when supporting a loved one with mental illness. The same is true for approaching your audience in science communication.

Today Mónica's work is a combination of communication, outreach, and administration. Every day is different for Mónica and she loves that! She works with scientists to tell stories about the importance of science in our everyday lives. She also works with teachers and students to help them use the science they are learning in the classroom to help their communities.

Mónica loves being able to serve Puerto Rico through science. Science can connect people to each other and the things that they love. She uses storytelling and her culture to help make science accessible so everyone can embrace the magic of science. It is important to her to give back to the Puerto Rican and Latinx communities she calls her own.

When she has free time, Mónica can still be found enjoying the outdoors or rocking out to Spanish and Afro-Caribbean music. Growing up on an island, she loves being at the beach. She and her husband like to travel, try new foods, and take their dogs - Ceci and Pati - for walks.

NICOLE ACEVEDO

Nicole Acevedo has dedicated her career to environmental health. Her mom is from Panama and her father is from Guatemala, so she was lucky to experience the nature and culture of different parts of the world at a young age. As a kid growing up in northern Virginia, Nicole wanted to be a veterinarian. Her dad was a doctor and she was always fascinated with anatomy - the part that grossed out her sisters.

She went to Virginia Tech to study Animal Science and fell in love with reproductive physiology. Physiology is the study of bodies and systems so this field focuses on the systems involved in making babies. She worked at the National Zoo on internship projects with cheetahs and black footed ferrets. Nicole learned that reproductive technologies can help save endangered species!

Nicole went on to graduate school after not getting into veterinary school. She knew there was a need for studying human fertility so she studied Biomedical Sciences. She looked at hormones and how they impact early development. Hormones are signaling molecules that are super important to our development and our daily lives. They impact things like appetite, sleep, fertility, and more. Nicole then moved to Massachusetts to work in a clinic that used reproductive technologies to help people have babies. She loved helping families grow but she knew that she wanted to learn more about what could be causing them to have trouble having a baby in the first place.

One day, Nicole picked up the book, *Our Stolen Future*, about how chemicals affect the environment and cause health problems in humans and animals. Nicole was fascinated with how it seemed to connect all her prior interests and experiences. On a whim, she reached out to one of the researchers referenced in the book.

Good thing she took that chance. The researcher asked Nicole to join her team as a postdoctoral researcher. Remember, sometimes you simply need to make the ask to help turn conception into reality. Nicole learned a lot about the different ways harmful chemicals affect our bodies and she also started to learn why it was so easy to be exposed to these chemicals in our water, air, food, and products.

Nicole knew she wanted to make real changes around how chemicals are used in everyday products. She joined a beauty company as their lead scientist to help them create products with safer chemicals. Now, Nicole is the owner of her own scientific consulting company that helps companies make safer products. She also works with organizations on issues like green chemistry, plastics pollution, and environmentally sustainable business practices.

Her work in academia and industry reminds her to give back, especially to communities of color that may have a higher lifetime risk of exposure to environmental toxins. She enjoys mentoring and sharing the message of how we can all play

 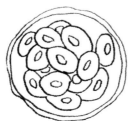 

a part in creating a cleaner and healthier world. Our choices as consumers have an impact on the planet and those choices are important.

Nicole says it's an exciting time to be a scientist and we agree. Her work is a reminder that we are all connected to the world around us.

PAMELA MCCAULEY

Pamela McCauley uses science and math to design safe products, equipment, and workplaces. Her area of expertise is called ergonomics and it uses a lot of biomechanics - the study of how the human body is built and how it moves.

Ergonomics is important because it helps us design products and spaces that keep us safe. For example, ergonomics is used to figure out the best size and most comfortable shape for airplanes seats. It's also used to find the best size of the text characters on your cell phone.

Pamela got her degrees - all three of them - in Industrial Engineering from the University of Oklahoma. Pamela is a professor at the University of Central Florida where she runs their Ergonomics Laboratory and her small business that provides expert witness and evaluation services. She also does a lot of work with girls and women in STEM.

Each week, Pamela's tasks are a little different. She spends several days a week teaching and doing her university research. At least two evenings a week are spent as an expert witness where she can speak on biomechanics and ergonomics issues. Part of her job is predicting risks. She can assess the risk of injury to people in different situations like car crashes or workplace accidents by using math and physics principles. Examining car accident cases can determine what factors were involved and how to make them safer in the future. And she travels on weekends at least three times a month for speaking engagements - many times encouraging women to think about science careers.

Pamela grew up in an Army family so she lived in a lot of different cities but identifies Oklahoma as home. Her parents were always a strong influence in her life and family is very important to her. She is one of six siblings - five girls and a boy. It was time spent at her grandmother's house that actually first started her interest in science. Her grandmother,

Mother Dear, lived out in the country and Pamela and her siblings would spend time there in the summers. They'd sleep outside so they could look up at the stars. Pamela started to learn about astronomy and it made her more curious about how the world worked. Her dad encouraged her to keep exploring her talents in math throughout school.

Two important things happened when Pamela was a teenager. One, she discovered the work of George Washington Carver - a black scientist who did a lot of work in agriculture, botany, and environmentalism. He also invented peanut butter! His work was an inspiration to Pamela to follow her STEM goals. More importantly, she had her daughter Annette.

Being a single mom as a teenager had it's challenges but Pamela had lots of family support and the drive to continue working toward her dreams of becoming a scientist. She loves how science can be used to understand, solve, and explain problems while coming up with solutions. She now uses her empowering story as encouragement to others.

PARISA TABRIZ

Parisa Tabriz works in cybersecurity, a field all about protecting computers and the data they store or access. She's known as the Security Princess and she works with hundreds of other security engineers at Google to keep us safe.

We've all become dependant on the web for information, connecting with other people, making purchases, getting work done, or just enjoying and sharing funny cat videos. Unfortunately, there are also online threats to users and the sensitive data they share with different services on the web. For example, online attackers try to make money off of passwords and credit card numbers they steal from websites that have security holes in them. Hackers are people that find security holes in software and computer systems.

Parisa is a white hat hacker, which means she uses her hacking skills for good - not evil. She works on Chrome - Google's web browser - to help keep over two billion people around the world safe as they use the web. Parisa and her team hack their own software to find security holes and then come up with solutions to patch those holes. They also build in layers of defense and help make security usable for all the people around the world that use Chrome on their laptops and phones.

One of her favorite things about working in cybersecurity is that she gets to work with people doing such a range of jobs - engineers, product managers, designers, lawyers, and more. She manages people located all over the world, and many of their projects are collaborations between many people with different skill sets working together toward a common goal. Security is at the cutting edge of technology, so she is always learning new things.

Parisa grew up in the suburbs of Chicago with her parents and brothers. As a young student, she liked math and science classes. Her dad was a doctor and her mom was a nurse, so she had a lot of exposure to science, but not to technology. In high school, she started using a computer for the first time to do research for school projects, play games, and talk to her friends. She went into computer science wanting to know more about how these incredible machines worked. Parisa discovered the security aspect of the field in college after one of her own sites got hacked, and she joined a computer club to learn more.

Being so connected for her job, Parisa likes to disconnect in her free time. She loves hiking and rock climbing in the mountains. Reconnecting with nature away from all the devices helps her have quiet time to think. She also likes doodling, photography, and spending time with her cats, Grace and Darwin.

Outside of her job at Google, Parisa spends time showing young people about computer science at hacking competitions. One of her role models was Grace Hopper, a pioneer in computer programming. Hopper was the person who coined the term "bug" in relation to computer programs. Parisa named her cat Grace in Hopper's honor. Perhaps there is a young lady ready to name her cat Security Princess as Parisa has become a role model herself.

PRASHA SARWATE DUTRA

Prasha Sarwate Dutra is an engineer with lots of enthusiasm for outreach.

Do you ever wonder how the extension cords you use are made? Engineers like Prasha make millions of feet of those rubber insulated cord cables everyday! In fact, that is exactly what she does at her day job as a Quality Manager for the largest wire and cable company in the world. They literally help connect the world as wires and cables are a key component of our modern technology and civilization. Plus the cables that Prasha makes are used in some super cool ways too, for example, they power roller coasters.

In college, Prasha studied Chemical and then Mechanical Engineering. She took aerospace engineering courses too since she was so inspired by Kalpana Chawla, an Indian astronaut on the Columbia Mission. Prasha applied that aerospace knowledge to build micro aerial vehicles (tiny flying machines) and design drones in international competitions. She loves that she is challenged daily at work and gets to be creative in solving problems that in turn touch the lives of so many people around the world.

Prasha was exposed to science as a child growing up in Delhi and Jaipur, India. Science is a popular subject in India and all children are encouraged early on to pursue those careers. Specifically for Prasha, her teachers, her parents, and her role model Kalpana Chawla also played an important role in fueling her curiosity and getting her into engineering.

The television show *How It's Made* was always on at the Sarwate household. Prasha remembers watching that show with her dad and that got her fascinated with manufacturing. A few years ago, Prasha lost her dad to pancreatic cancer. Dealing with his death was difficult for her and it brought on a challenging period of depression. Prasha has dealt with depression before, like when she moved 9,000 miles away from home to attend graduate school in the United States. She has used those experiences to rediscover herself and the importance of self love. She has recently started CrossFit to stay healthy and loves it. Practicing meditation also helps her stay present in her life and she carries that everywhere - work, outreach, and fun!

One of Prasha's latest activities is reading more books. She's challenged herself to read 100 books this year. She also likes to spend time connecting with family - both in person and virtually. Her side of the family lives in India. Her husband's family lives right down the road. She feels very grateful to have such a funny and supportive family group.

Prasha strongly believes that women can and will change the world. She hopes to inspire more women to join and stay in STEM through her podcast, *Her STEM Story*. The podcast shares the stories of real women in STEM and also fuels her own motivation. In each episode, she interviews a STEM related scientist or someone who works in a STEM organization about their work, their side projects, and their journeys. She too, is a woman connecting and changing the world.

SARAH MYHRE

Sarah Myhre grew up in Seattle, Washington enjoying all the glory of the Northwest from the North Cascade Mountains to the beaches of the San Juan Archipelago.

When she was in college, Sarah realized that she could get paid to travel and SCUBA dive if she studied marine biology. So she dove into the adventures of seeing the world - above and underwater. Before she went on to graduate school, she spent over 1,000 hours underwater as a scientific diver collecting health data on coral reef populations.

Today, she is back in Washington State working as a research associate studying climate and ocean science. The past is the key to the future and this is true with the oceans too. Sarah studies the ocean sediments to reconstruct the history of the planet to help understand our world today. There are clues in the dirt, sand, and clay at the bottom of ocean basins that she uses to learn about how the atmosphere operated in the past.

Each work week looks different for Sarah - this is true for many scientists and it keeps things exciting! Her ocean sediment research takes time looking at samples under microscopes and analyzing data at a computer. Her science communication work takes time writing manuscripts for scientific journals as well as articles for broader audiences. Sarah has also been a guest on podcasts and radio programs talking about climate science. By putting herself out there as a resource she is helping more people understand why her type of science is important to everyone and to the planet.

Sarah's latest adventure in science communication is entering the political arena. She became involved with a group of women scientists in Washington State who organize to bring science issues to local government officials. Her past experience in giving public talks about science in her communities was good training for entering the policy arena. The messages have to be super clear and concise since politicians are very busy and may not have a scientific background. Sarah also has to keep her audience in mind by making sure to address the issue in a way that her listener will relate to and find important to their own lives, work, and families. She combines all her experience in science, communication, and policy in her work at the Rowan Institute where she consults with others trying to make positive and inclusive change in STEM.

Family is super important to Sarah. She has a toddler who she says is helping her be a better person. Being a parent pushes her to be efficient, organized, and thoughtful - all things that come in handy as a scientist too. The two of them do a lot of outdoor activities with Sarah's family - her two brothers and her parents. Her parents met at college getting their computer science degrees and now her brothers work in technology too. You could say they're a science family and having that support helped Sarah in her journey becoming a scientist.

Sarah loves scientists and how they show up every day for the pursuit of truth and knowledge. She knows that we need the culture of science to be inclusive, diverse, and equitable. Her work in science communication and policy is how she is contributing to making that a reality.

SUNSHINE MENEZES

Sunshine Menezes is a super science communicator who studied zoology and oceanography.

Oceanographers are interdisciplinary, meaning they use many different fields of science in their work. They use biology, chemistry, geology, and physics to learn about the oceans and shoreline. Sunshine studied plankton, some of the smallest plants and animals in an estuary - a coastal environment where salty ocean water mixes with freshwater. She took thousands of water samples to figure out where these tiny organisms liked to live in the estuary. Studying the diversity of those tiny plankton helps scientists understand how estuaries might respond to environmental changes.

Today, Sunshine runs the Metcalf Institute at the University of Rhode Island. Metcalf Institute helps people understand the science related to environmental issues, from clean air and water to climate change. The Institute trains journalists and scientists how to explain science more clearly and show how science impacts our lives. Since so many people get information on environmental issues from the news, helping journalists and scientists communicate more clearly helps all of us get more accurate information. No work week is the same for Sunshine. She organizes Institute workshops, writes grants to fund those projects, travels to run training programs, gives public talks about science communication, and teaches college classes on how to get more people interested and involved in science.

Growing up in northern Michigan, Sunshine had unlimited space to play outside. She made mud pies in her yard, took long bike rides around the rural farms, and explored the woods. Sunshine's dad built their house and they didn't have electricity until she was in fifth grade! Both her parents were big role models for Sunshine and taught her how important it is to stand up for yourself. Her mom was a nurse and then a nursing teacher. Her dad went to law school when Sunshine was in middle school and became a lawyer and eventually a judge.

Most of her extended family lived in Florida so Sunshine often visited the ocean. Her mom says she was interested in ocean animals as a very young child. Sunshine remembers poking all the dead animals that washed up on the beach to see how they looked on the inside. That curiosity led her to study biology. Science and research didn't come easy to Sunshine, but she thinks it's important to do things that push her to grow. One thing that helped her grow as a scientist is that she loves asking questions - really the core of what scientists do.

Asking questions played a big role in an unexpected challenge for Sunshine. When she was 37 years old, she was diagnosed with cancer for the first time. She had surgery, radiation, and chemotherapy. Today, she is a healthy, three-time cancer survivor. Cancer is scary and she knows she could become sick again. Sunshine says that being a scientist helped her face her sickness because she was able to approach it with lots of questions, research the effects of the disease, and understand her treatment options.

Hi, I'm a copepod.

Outside of work, Sunshine loves to sing and write original songs like those on her album *Shiner and Her Lucky Pennies*. She also enjoys gardening, traveling, learning new languages, board games, and taking walks with her husband and dog, Kit.

SYLVIA ACEVEDO

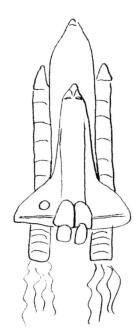

Meet lifelong Girl Scout and industrial engineer, Sylvia Acevedo. Industrial engineers, also called systems engineers, are problem solvers that focus on how people can get the most out of experiences at work and at play. Their work can range from making theme parks enjoyable for visitors to making manufacturing plants super efficient so that they can make more products faster and safer.

Sylvia was born in South Dakota and grew up in New Mexico. Her dad was in the Army and worked as an analytical chemist. Her mom took care of the household and made sure Sylvia and her three siblings were focusing on school. She was always interested in how the world around her worked and wanted to understand all she could. As a kid, she participated in Head Start and Girl Scouts - two programs dear to her heart and core to her advocacy today.

She remembers a Girl Scout troop leader encouraging her to earn her science badge. The troop leader caught young Sylvia looking up at the stars and pushed her to learn more about astronomy. To earn that badge, Sylvia built a rocket that she launched into the New Mexico night sky and the rest is history.

Sylvia studied at New Mexico State University and then Stanford University for graduate school. She was studying STEM at a time when not many women were going into those fields and even fewer women were becoming engineers. At her first job as a field test engineer, the company didn't even have a women's restroom when she started. She was inspired by powerful women from history who saw systems that needed fixing and took matters into their own hands to solve problems. Sylvia knew she'd have challenges to overcome and worked harder than those around her to show them she belonged. The challenges didn't keep her down, she literally became a rocket scientist!

She spent many years working in laboratories where she used science, mathematics, and engineering methods to analyze systems to find ways to make them work better. One of her first projects was figuring out how equipment would work in outer space. She had to think about issues that might come up in that environment - no gravity, radiation, extreme heat, and magnetic fields - and anticipate how they could affect the equipment. This process involved hours and hours of making and solving complex algorithms. An algorithm is a set of specific steps and rules for solving a math problem, oftentimes algorithms are performed by computers since the steps are repeated so many times.

Now, Sylvia is the head of the Girl Scouts of America. Her mission is to inspire more students to study science, technology, engineering, and math. She's even written a book about it. The best part of the job for her is meeting young girls who she knows will be the leaders and scientists of the future.

Outside of work, Sylvia loves to spend time with friends and family - especially at the beach and on hikes with their family dog, Xochitl, an Australian Kelpie mix.

TIFFANY LYLE

Tiffany Lyle studies veterinary anatomic pathology - that means she studies disease in animals from rodents to elephants. Veterinary science is much more than the field that helps your household pets. Veterinarians are also critical players in the research world.

Growing up with accountant parents in Indianapolis and Atlanta, Tiffany didn't have her first real science experience until tenth grade. She participated in the science fair with a project about eye diseases and was encouraged to continue research in a biochemistry lab as an undergraduate at the University of Georgia. For graduate work, Tiffany completed research at the National Cancer Institute in Maryland and received her doctorate from Purdue University. She also has a doctorate of Veterinary Medicine from the University of Georgia. She's a double doctor!

She speaks highly of her mentors and their patience, graciousness, and true caring for people. Today, she brings those invaluable practices to her own laboratory making sure to take responsibility for the people who work with her and respecting their voices and opinions.

Tiffany is a professor at Purdue University in Indiana. She researches a section of the body called the blood-brain barrier which protects the brain from toxins, hazardous chemicals, and dangerous cells. Her research is looking at what happens in the blood-brain barrier after cancer cells grow inside the brain. Since the barrier is designed to keep things out, it is difficult to design helpful drugs that can get in. Tiffany develops animals models to look at these diseases and how we might be able to use those tools to help people.

Understanding how cancer cells move throughout the body and the brain is critical to developing treatments. Research like Tiffany's helps find ways to not only shrink cancerous tumors but also to potentially stop them from migrating to other parts of the body.

She teaches future veterinarians and also does a lot of science outreach work in her community. A regular week for Tiffany is a full one. She prepares lectures, connects with her research lab, and oversees a group of pathology technicians. She also has to write a lot for grants that help fund her research and travels for conferences to meet other scientists from all over the world. Of course there are also meetings each day with students and collaborators who are working together to learn more about the brain.

One of her favorite things about her job as a professor is celebrating her students' successes. The sheer joy of discovery is an amazing thing and she gets to teach students about that aspect of research. Sure, sometimes research can feel unrewarding, but then there are the moments where YOU are the first person to know something new and that is super cool.

Tiffany prioritizes her health and family time by getting up early to exercise and spending time after work with her husband and son, the Georges. Her family spends time together by the pool and she loves cooking, running half marathons, and reading. Maya Angelou is one of her favorite authors.

TRACY CHOU

Tracy Chou is a software engineer. Computers are amazing machines that can follow instructions well and super fast. People, like Tracy, have to design and write those instructions, often referred to as code, and the systems they follow. She has worked at Quora and Pinterest doing this type of work. Her favorite thing about being a software engineer is that she gets to build things that never existed before that millions of people use every day.

What exactly does a software engineer do? Tracy spends about half her time writing and revising code and making sure existing systems work well. Just like traditional writing, going back to review the code is important. This helps you find errors and areas where you can improve the writing. Tracy is often a tech lead in software projects so she spends the rest of her time doing more management like prioritizing tasks and figuring out the plan to get those things done. Tasks can be people related like collaborating with her teammates or tech related like deciding how to set up data storage.

Growing up in California's San Francisco Bay Area, Tracy was in a great place to grow up learning about technology. Tracy enjoyed and was good at math and science as a kid. She went on to Stanford University to study Electrical Engineering as an undergraduate and then got a graduate degree in Computer Science. You could say coding was in her blood. Her parents are also both software engineers and so is her sister.

Even with all that STEM support at home, Tracy is open about some of the challenges she faced as a woman in technology, including going through imposter syndrome. Imposter syndrome makes it hard to recognize our own accomplishments and talents in an environment that doesn't feel like it is made for us. It's pretty common for women in STEM and Tracy uses that experience to support other women and also to push companies to be accountable with their people practices. Tracy knows that it can still be difficult for women - not because of lack of interest or ability - but because of the slow chipping away of confidence that can happen in tech culture. She wants to change that.

Working in such a data-driven industry, she was curious to know the actual numbers behind the diversity issues we hear so much about in tech. She launched an initiative called *Where Are The Numbers* that asked tech companies to fill out a survey about their employee diversity data. This project launched her into the public eye as a diversity activist and hundreds of companies participated. As suspected, the numbers of women and people of color were low. However, now there is actually baseline data and that helps us do two things. One, it helps us actually identify the problem - the first step in coming to solutions. And two, it will allow us to track progress to know if the solutions in place are truly making a difference.

The numbers help us find ways to help people - via tech products and inclusion in tech careers. When Tracy isn't crunching the numbers, you can find her hiking, doing yoga, and spending time with the important people in her life.

TRISHA WALSH

Trisha Walsh works in computer information systems or information technology. Businesses need integrated computer systems in order to function. This is true for all businesses including technology, banking, insurance, healthcare, and retail. Information technology helps run business operations, find new customers, and make communications easy.

Today, Trisha is the Director of Information Technology at Yelp! Yelp is an application that provides reviews of restaurants, hotels, and other services. Her week is a lot of managing people. She is at a point in her career where most of her work is managing the managers of teams in her department. Trisha works with the managers to make sure they are reaching goals, guiding their professional development, and listening. By listening closely, she can better understand what any and all members of the team may need for support to get their job done and done well. She loves the positive and supportive work environment.

Trisha says she got interested in computer science by accident. After high school, she didn't go to college right away. It was after graduating high school that she found a mentor in one of her high school math teachers who was also an antique shop owner. One afternoon in the shop, her teacher asked Trisha why she wasn't in school especially since she had done so well in her advanced math classes. When her mentor found out Trisha didn't have the money to go to college she sent her to the high school guidance office for help in understanding the financial aid process.

She got her undergraduate degree in sociology and social work and went on to work at a large financial institution. It was on the job where she learned she could fix issues that came up with printers and fax machines. She quickly learned documentation applications and found she was good at troubleshooting and solving problems. So she moved into a job doing application testing (also known as quality assurance) and went back to school to study technology, specifically computer information systems.

Growing up in Massachusetts, Trisha learned a lot from her parents including the importance of family. Now, she calls a group of 28 beautiful people her immediate family including siblings, spouses, nieces and nephews, step kids, and grandchildren. Trisha loves spending time with her wife and their large family. They enjoy reading, watching movies, and hiking in nature. Even though her parents are no longer here, their lessons continue to guide Trisha to give back both in her work and her life outside it.

Outside of work, Trisha has several side projects focusing on outreach. One project is with a non-profit specializing in healthcare for the whole person, meaning healthcare that includes primary care, mental health, substance use issues, and social support for people re-entering the workforce after prison. The second project is a podcast she started called *Techgrlz*. *Techgrlz* brings the stories of women in technology to life to inspire more young women to pursue technology careers at any and all points of their journey, like Trisha.

RESOURCES

This list is by no means comprehensive but it is a good starting point if you want to learn more about some of the STEM fields or organizations mentioned in this book or learn more about what is out there for girls and underrepresented groups in STEM and STEAM.

ORGANIZATIONS + PROGRAMS

500 Women Scientists
https://500womenscientists.org
This organization works to make science open, inclusive, and accessible by providing advocacy in leadership, diversity, and public engagement.

A Mighty Girl
www.amightygirl.com
A collection of books, music, movies, and more resources dedicated to raising confident, smart, and brave girls. Lots of STEM suggestions.

African American Women in Physics
www.aawip.com
The AAWIP page celebrates black women with physics degrees working in a wide variety of STEM fields. You can learn more about them and follow their blog for women in physics news.

AI 4 ALL
www.ai-4-all.org
AI 4 ALL educates the next generation of artificial intelligence technologists, thinkers, and leaders through summer programs located at universities across the country.

Alaska Native Science and Engineering Program
www.ansep.net
ANSEP supports Alaska Native students in STEM careers by providing support and programming starting in middle school continuing through graduate work.

American Indian Science and Engineering Society
www.aises.org
AISES is focused on increasing American Indians, Alaska Natives, Native Hawaiians, Pacific Islanders, First Nations, and other indigenous peoples of North America in STEM careers. They offer a variety of resources including scholarships and internships.

Association for Women in Science
www.awis.org
AWIS supports the advancement of women in STEM through a national research platform, advocacy, and more.

Biomedical Engineering Society
www.bmes.org
The BMES is the professional home for biomedical engineering and bioengineering. They serve as a forum for publications, scientific meetings, and mentor opportunities.

Black Girls Code
www.blackgirlscode.com
BGC introduces computer coding to young girls of color through workshops and school programs. See their website for events and volunteer (even non-tech options) opportunities.

Black Women in Science and Engineering
www.bwiseusa.org
BWISE is a group where black women in STEM can connect and empower each other.

Broadcom MASTERS
https://student.societyforscience.org/broadcom-masters
A prestigious national science competition for middle school students.

CienciaPR
www.cienciapr.org
Our Super Cool Scientist, Mónica Feliú-Mójer, works for this organization that shares a collection of bilingual stories of Puerto Rican and Lantinx scientists including specific blogs on women, health sciences, and more. The site also includes resource sections for educators and K-12 students.

Circuit Breaker Labs
https://circuitbreakerlabs.myshopify.com
Our Super Cool Scientist, Amanda Preske, makes amazing science-themed jewelry from recycled computer parts and shares the stories of women in STEM.

Enhancing Diversity in Graduate Education
www.edgeforwomen.org
The EDGE Program provides programming and support to women completing doctorate programs in mathematics. They offer summer sessions, conferences, travel for research, mentoring activities, and more.

First Robotics
www.firstinspires.org
First Robotics inspires young people to get interested in STEM by building technology skills and teamwork through robotics competitions and other challenges.

Frontiers of Flight Museum
www.flightmuseum.com
Our Super Cool Scientist, Alicia Morgan, oversees the education programs at this museum that let you explore the stories of aviation and space flight.

Galaxy Zoo
www.zooniverse.org/projects/zookeeper/galaxy-zoo
Learn about galaxies and how to identify them at this super cool website.

Girl Scouts of America
www.girlscouts.org
Our Super Cool Scientist, Sylvia Acevedo, leads this organization. Learn more about what Girls Scouts offers and read great articles on how to empower our young women to become leaders. Participating in Girl Scouts helps girls develop a strong sense of self, seek challenges and learn from setbacks, display positive values, form and maintain healthy relationships, and identify and solve problems in the community.

Girl STEM Stars

www.girlstemstars.org

Our Super Cool Scientist, Camille Eddy, is on the board of this academy dedicated to advancing girls of color from underserved communities in science, technology, engineering, and mathematics. They provide mentoring and special field trips to expose young girls to the opportunities and role models in STEM.

Girls Auto Clinic

http://girlsautoclinic.com

Our Super Cool Scientist and Engineer, Patrice Banks, created the Girls Auto Clinic workshops, glove box guide, and first auto repair shop catered to women complete with beauty bar.

Girls Who Code

https://girlswhocode.com

Girls Who Code wants to close the gender gap in technology. They offer programs for young women to learn coding schools over the summer and school year. If there is not a current chapter near you, they also provide resources to start your own.

How It's Made

www.sciencechannel.com/tv-shows/how-its-made

Our Super Cool Scientist, Prasha Sarwate Dutra, loved watching this show with her dad growing up. It shows viewers how all sorts of products are created, giving an insider look into the manufacturing and engineering processes involved.

Human Factors and Ergonomics Society

www.hfes.org

HFES hosts a range of educational resources, a career center, expert witness information, and more for students and professionals interested in ergonomics.

iBiology

www.sciencecommunicationlab.org/backgroundtobreakthrough

Our Super Cool Scientist, Mónica I. Feliú-Mójer, produces this series of science videos for non-scientists.

Infosec Rocks

https://sites.google.com/site/infosecrocks

Infosec Rocks is a collection of activities and resources for anyone who wants to learn about cybersecurity.

Institute for Electrical and Electronics Engineers

www.ieee.org

IEEE is a technical professional organization providing resources to people interested in technology including conferences, education, and career information.

Institute of Industrial & Systems Engineers

www.iise.org

IISE is a professional society for industrial engineers that provides leadership, training, conferences, a career center, and more.

Kepler Telescope

www.nasa.gov/mission_pages/kepler/overview/index.html

Learn more about the Kepler Telescope mission including an up to date count of how many planets the telescope has helped discover.

LabCandy

http://labcandy.com

Our Super Cool Scientist, Olivia Pavco-Giaccia, started LabCandy to get young girls interested in science through educational kits with jeweled goggles, colorful lab coats, and interactive story books.

Latinas in STEM

www.latinasinstem.com

Our Super Cool Scientist, Noramay Cadena, co-founded Latinas in STEM to empower Latinas to thrive and advance in STEM fields. Visit their website for professional development, K-12 outreach, and college student support resources. Resources for parents too!

Make in LA

http://makeinla.com

Our Super Cool Scientist, Noramay Cadena, works with LA hardware startup companies to help local entrepreneurial leaders build solid product and business foundations through providing facilities, mentorship, and intensive skills training.

Maker Faire

https://makerfaire.com

The Maker Faire festival is a family-friendly event that highlights creativity and invention. There are lots of neat STEM-themed products like the circuit board jewelry made by Super Cool Scientist Amanda Preske.

March for Science

www.marchforscience.com

The March for Science movement is a group working in science advocacy. Go to their website to learn more about what they are doing near you.

Math Mamas Facebook Group

Our Super Cool Scientist, Emille Davie Lawrence, started this closed Facebook group for other mathematicians who are also mothers. Search for "Math Mamas" and ask to join the group.

Metcalf Institute for Marine and Environmental Reporting

http://metcalfinstitute.org

Our Super Cool Scientist, Sunshine Menezes, runs the Metcalf Institute which helps journalists and scientists better communicate science to a broad range of audiences through programs, workshops, and an intensive summer institute.

Minorities Striving and Pursuing Higher Degrees of Success

www.msphds.org

Our Super Cool Scientist, Ashanti Johnson, founded MS PHD's. They work to increase the participation of underrepresented minorities in earth system science through professional development, networking, and mentoring.

Mission Blue

www.mission-blue.org

Our Super Cool Scientist, Sylvia Earle, started Mission Blue to ignite public support for the blue heart of our planet. They advocate for the creation of marine protected areas worldwide.

Mostly Microbes Blog

www.mostlymicrobes.com

Our Super Cool Scientist, Anne Estes, runs this website that has lots of stories and resources about microbes.

myFOSSIL

www.myfossil.org

The FOSSIL project aims to promote paleontology outreach, education, and collaboration through an online community of resources and support.

National Aeronautics and Space Administration

www.nasa.gov

Learn more about NASA missions, astronauts, and a full range of space science. They have amazing photo galleries as well as educational tools like podcasts.

National Girls Collaborative Project

https://ngcproject.org

The NGCP works to unite organizations around the country that encourage girls to pursue STEM. Visit their website for a program directory, events, and what's going on in your state.

National Math + Science Initiative

www.nms.org

NMSI works to improve student performance in STEM by providing professional development and college readiness resources like SAT vocabulary practice.

National Museum of Mathematics (MOMATH)

http://momath.org

Lots of hands-on exhibits, galleries, and programs about the wonders of mathematics.

National Society of Black Physicists

http://nsbp.org

The NSBP promotes the professional well-being and success of African American physicists and physics students. See their website for job boards, conference information, and more.

National Society of Hispanic MBAs

www.nshmba.org

The NSHMBA has a range of programs and services for networking, job opportunities, and events including an annual conference.

National Society of Hispanic Physicists

www.hispanicphysicists.org

NSHP celebrates Hispanic physicists and supports their professional work. Visit the website for meetings, publications, career services, and more.

Natives in STEM

www.nativesinstem.org

The Natives in STEM project features stories of Native professionals to increase the visibility of Native people in the STEM community in hopes to inspire others.

NIDA for Teens

https://teens.drugabuse.gov

A blog specifically for teens focused on neuroscience and drug addiction awareness.

NOAA Corps

www.omao.noaa.gov/learn/noaa-commissioned-officer-corps

The NOAA Commissioned Officer Corps is one of the United State's uniformed services that combines science expertise with service. Officers like our Super Cool Scientist, Cherisa Friedlander, serve on land, in the air, and on the sea.

Planet Hunters

www.planethunters.org

Learn all about the planets and even discover new ones!

Real Impact Center, Inc.

www.realimpactcenter.com

Various education programs for girls in STEM, based in Georgia.

Rochester Museum and Science Center

www.rmsc.org

This RMSC promotes community interest and understanding of science and technology through hands-on exhibits, special programs, a planetarium, and more.

Rowan Institute

www.rowaninstitute.org

Our Super Cool Scientist, Sarah Myhre, founded the Rowan Institute to teach STEM professionals to better communicate complex information, to train public leaders with a scholarship and social justice lens, and more.

Science Trek on PBS

www.pbs.org/show/science-trek

Public Broadcasting Service is great for educational content and Science Trek is the super cool science web and broadcast project for children to learn more about STEM and have their STEM questions answered by experts.

Smore Magazine,

www.smoremagazine.com

A magazine for young people, especially girls, to encourage their curiosity and more.

So, you want to work in security?

https://medium.freecodecamp.org/so-you-want-to-work-in-security-bc6c10157d23

Our Super Cool Scientist, Parisa Tabriz, gives a bunch of information and links to everything security related in this article.

Society for Science & the Public

www.societyforscience.org

This group puts on the Intel International Science and Engineering Fair and other programs to get people of all ages involved in STEM activities.

Society of Hispanic Professional Engineers

www.shpe.org

The SHPE offers events, programs, scholarships, and more for K-12, college students, and professionals. Visit their website for more information.

Society of Women Engineers

http://societyofwomenengineers.swe.org

SWE provides scholarships, K-12 outreach opportunities, and networking for women engineers worldwide. Visit their website for more information, a career portal, a magazine, and more.

Tested

www.tested.com

Adam Savage takes on all sorts of science and technology topics in Tested. Check out their videos and podcast for more information on airplanes, robots, and so much more.

Texas Instruments Education

https://education.ti.com

TI supports STEM education. Resources include test prep tools, lessons, and the "STEM behind" series that explores the topics in things like sports, movies, and health.

The Urban Scientist

https://blogs.scientificamerican.com/urban-scientist

Our Super Cool Scientist, Danielle N. Lee, blogs about her research and outreach initiatives.

USA Science and Engineering Festival

https://usasciencefestival.org

This is a huge STEM festival complete with exhibits, a career fair, and more.

Vi Hart's YouTube Channel

www.youtube.com/user/Vihart

Fun math-themed videos by Vi Hart, daughter of George Hart who makes famous mathematical and geometrical sculptures.

Violet STEAM Project

www.violetsteamproject.com

Our honorary Super Cool Scientist, Violet, started the Violet STEAM Project to celebrate women in STEM. She and her mom worked with an artist to create posters and shirts with the images of some amazing women in STEM (like Grace Hopper - it is a Violet STEAM Project poster on the wall of Parisa Tabriz's illustration in this book).

Where Are the Numbers?

https://medium.com/@triketora/where-are-the-numbers-cb997a57252

Our Super Cool Scientist, Tracy Chou, started this initiative to learn more about the baseline diversity numbers at tech companies.

Zooniverse

www.zooniverse.org

Zooniverse features numerous projects covering many different science fields. It teaches any person how to participate and help scientists with a variety of research projects.

PODCASTS

Femmes of STEM

www.femmesofstem.com

Our Super Cool Scientist, Michelle Barboza, hosts the Femmes of STEM podcast highlighting historical women in science with special STEM guests from today.

Her STEM Story

https://herstemstory.com

Our Super Cool Scientist, Prasha Sarwate Dutra, interviews real women in STEM in this podcast. The interviews dive into their work, their outreach, and their journeys. In addition the podcast there is also an active blog, a book club, and more to check out.

PhDivas

https://soundcloud.com/phdivas

Our Super Cool Scientist, Liz Wayne, and Super Cool Humanities counterpart, Xine Yao, talk about academia, culture, and social justice in this podcast that recognizes and supports women in higher education.

TechGrlz

http://techgrlz.com

Our Super Cool Scientist, Trisha Walsh, shares stories of women and girls in technology with the aim of getting more women exposed to technology careers.

BOOKS

Girls Auto Clinic Glove Box Guide
By: Our Super Cool Scientist Patrice Banks

Good Night Stories for Rebel Girls
By: Elena Favilli and Francesca Cavallo

Headstrong: 52 Women Who Changed Science and the World
By: Rachel Swaby

Her STEM Career: Adventures of 51 Remarkable Women
By: Diane Propsner

Hidden Figures
By: Margot Lee Shetterly

Lab Girl
By: Hope Jahren

Path to the Stars: My Journey from Girl Scout to Rocket Scientist
By: Our Super Cool Scientist Sylvia Acevedo

Pink Boots and a Machete
By: Our Super Cool Scientist Mireya Mayor, Ph.D.

Rad American Women A-Z: Rebels, Trailblazers, and Visionaries who Shaped Our History...and Our Future
By: Kate Schatz

STEAM+ Arts Integration: Insights and Practical Applications
By: Jacqueline Cofield with chapter by our Super Cool Scientist Alicia Morgan

Winners Don't Quit: Today They Call Me Doctor
By: Our Super Cool Scientist Pamela McCauley Bush, Ph.D.

Women's Adventures in Science Collection via National Academies Press
www.nap.edu/collection/54/womens-adventures-in-science

Women in Science: 50 Fearless Pioneers Who Changed the World
By: Rachel Ignotofsky

Women Who Don't Wait in Line: Break the Mold, Lead the Way
By: Reshma Saujani

GLOSSARY

3-D Space - also known as three-dimensional space, has three elements - height, width, and depth.

Ableism - the prejudice and discrimination of people with disabilities.

Advocate - someone who actively supports, promotes, and defends a cause such as inclusion in STEM.

Aeronautical Engineering - the engineering related to flight and the operation of aircraft.

Aerospace - an area focusing on airplanes and spacecraft.

Agriculture - the science of farming which includes the use of land and breeding of plants and animals to provide food and other products.

Algorithm - a set of specific steps and rules for solving a math problem, oftentimes algorithms are performed with computers since the steps are repeated so many times.

Amygdala - the part of the brain that controls emotion in humans and plays an important role in risk-taking behavior.

Analytical Aquatic Chemistry - the study of the properties of water, particularly the chemical components.

Analytical Chemistry - a type of chemistry that separates, identifies, and quantifies matter breaking it down in order to understand its components.

Anatomy - the part of biology that studies the structures of organisms and their parts. For example, in humans, our bones and muscles.

Animal Behavior - the study of how an animal's physical body and components connects to its behavior.

Animal Science - the study of the biology of animals.

Anthropology - the study of humans over time and space including their social interactions and culture.

Artificial Intelligence - the study of the intelligent behavior that machines or computers can exhibit that mimic natural intelligence in humans. Intelligent behaviors include things such as learning and problem solving.

Astronaut - someone who travels into outer space, beyond Earth's atmosphere.

Astronomy - the study of objects in outer space, also known as celestial objects.

Astrophysics - the type of astronomy that focuses on the stars and the physics behind understanding celestial observations.

Atoms - a tiny particle of an element that can exist by itself or in combination with other atoms in a molecule.

Atmosphere - the layer of gases surrounding a planet.

Autism - a range of developmental conditions characterized by challenges with social skills and communication.

Aquanaut - someone who lives underwater for a period of time in an underwater environment.

Bachelor's - a type of college degree, usually received first or early in someone's college experience, also known as an Undergraduate degree.

Baleen - the filter-feeding system found in baleen whales that looks sort of like their teeth.

Behavioral Health - includes both mental health and substance use; covers prevention, intervention, treatment, and recovery services.

Biochemistry - a combination of biology and chemistry, the study of the chemical processes that take place within living things.

Biodegradable - something that breaks down naturally in the environment; biodegradable materials are important for the environment because they prevent additional trash from going into landfills.

Biofilm - a thin layer of bacteria that sticks to a surface like the bottom of a boat.

Bioinformatics - the science of analyzing and making sense of large amounts of biological data like genomes; the computer-science side of biology.

Biology - the study of living things including animals, insects, plants, and microbes.

Biomechanics - the study of how the human body is built and how it moves.

Biomedical Sciences - the set of sciences that relate to healthcare and medicine.

Biotechnology - technology that uses biological processes.

Blood-Brain Barrier - a thin layer around the brain that protects the brain from toxins, hazardous chemicals, and dangerous cells.

Botany - the study of plants.

Chemical Engineering - the engineering that focuses on the principles of chemistry to design large scale processes that convert chemicals, raw materials, and more into useful products.

Chemical - a substance or compound that has been purified so it can not be broken down into smaller components without breaking molecular bonds.

Chemistry - the study of matter (solid, liquid, gas, plasma) and the properties of matter.

Civil Engineering - the engineering that deals with the design of environments, natural or human-made.

Climate Change - a change in global climate patterns due to increased greenhouse gas emissions that has effects on sea level, severe weather, crop cycles, animal migrations, and more.

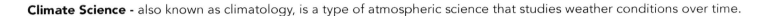

Climate Science - also known as climatology, is a type of atmospheric science that studies weather conditions over time.

Code - the programming language used to create the instructions that run computers systems, there are many different types of coding languages.

Coding - the writing of computer software to design and create systems like phone applications; sometimes called Programming or Computer Programming.

Cognitive Science - the study of the mind and how it works.

Computational Biology - the science of analyzing and making sense of large amounts of biological data like genomes.

Computational Model - using a computer to model complex systems using mathematics, physics, and sometimes bioinformatics.

Computer Science - the study of automating sets of instructions performed in sequence (algorithms) that can be processed in large amounts.

Conservation - the protection of the planet's natural resources including plants and animals through policy like marine protected areas and National Parks.

Cultural Bias - the tendency to judge another person based on their culture or background.

Crystalline Structures - a structure with atoms or molecules that lines up in an ordered arrangement called a crystal.

Cybersecurity - a technology field that focuses on protecting computers and they data they store or access.

Dark Matter - a mysterious type of matter that makes up over a quarter of the universe.

Deoxyribonucleic Acid or DNA - the genetic code that serves as the foundation of life. DNA is a molecule made up of 4 base units and it has a double helix structure like two staircases twisting around each other with railings on the outside. This unique structure, plus the fact that each unit has a specific pairing, allows DNA to act as a code.

Design Thinking - a creative process used for problem solving and innovation that uses the practices of design such as empathy, idea generation, and experimentation.

Diversity - when talking about ecology, diversity is the makeup of all the different types of organisms in an environment and how many of each type.

Doctorate - an academic degree that puts someone at the top of their field, sometimes called a terminal degree, and qualifies them to teach at university.

Drone - also known as an unmanned aerial vehicle, an aircraft without a pilot onboard, the crafts are controlled remotely.

Ecology - the study of how all living things and their environments interact.

Ehlers-Danlos Syndrome - also known as EDS, a rare genetic condition that inhibits mobility and causes chronic pain.

Element - a chemical substance that can not be broken down into simpler substances.

Endangered Species - a species of organism (animal, plant, etc.) that has so few individuals left in their population that the species is at risk of going extinct.

Engineering - the application of mathematics and science to design useful products and systems; there are many types of engineering such as Aerospace, Chemical, and Mechanical.

Entrepreneur - someone who starts their own business.

Environmental Engineering - the engineering related to our physical environment including law, air pollution, water issues, and public health.

Environmental Health - a component of public health that focuses on the parts of the natural world and created environment that affect human health.

Environmentalism - a movement concerned with protecting the natural world.

Epilepsy - a neurological disorder that can cause seizures of varying intensity.

Ergonomics - a field that focuses on the design of products and spaces that keep us safe using the principles of biomechanics or the study of how the human body is built and how it moves.

Estuary - a coastal environment where salty ocean water mixes with freshwater.

Expert Witness - a professional who is a specialist in their field so much so that they are considered an expert by judges in court cases.

Extragalactic - referring to outside of the Milky Way galaxy in astronomy.

Evolutionary Biology - the study of how living things adapt to new situations and environments over time.

Fertility - the ability to produce young or offspring.

Gamma Rays - a type of ray that has the most energy, and smallest wavelength, in the electromagnetic spectrum.

Genetics - the study of an organism's DNA sequence and genes.

Geology - the science of Earth's physical structure and composition including the study of things like rocks, minerals, volcanoes, earthquakes, and more.

Geometry - the field of mathematics that deals with shape, size, position of figures, and the properties of space.

Graduate Advisor - in graduate school, students have a faculty advisor who helps guide them through their research and professional path.

Graduate School or Graduate Studies - a more specialized type of academic degree, usually a master's or a doctorate.

Gravity - the force that draws objects, like planets, towards each other.

Green Chemistry - also known as sustainable chemistry, focuses on designing processes and products that limit the use of natural resources and keep hazardous chemicals to a minimum.

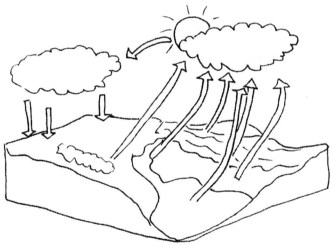

Habitability - being able to support life.

Hacker - a computer expert that uses their technical knowledge and skills to solve a problem relating to computer systems.

Hormones - signaling molecules that are important to our development and daily lives affecting things like appetite, sleep, and fertility.

Host - any kind of organism - humans, other animals, plants, insects - in or on which microbes live.

Hydrology - the study of water's properties, especially how it moves in relation to land.

Imposter Syndrome - when someone finds it challenging to recognize their own accomplishments and talents in a certain environment because they don't feel that environment is made for them to be active participants.

Industrial Engineering - the engineering related to getting the most out of complex systems, including things like saving time, money, and work hours. Sometimes called Systems Engineering.

Interdisciplinary - areas of study that use many different fields of STEM.

Lantinx - a Latin American person, often used as an inclusive, gender neutral or non-binary identifier.

Linear Algebra - the study of solving linear equations.

Magnetic Field - the magnetic effect of electric currents and magnetic materials, specified by both a direction and a strength so it is a type of vector field.

Manufacturing - the production of products using labor, machines, automation, and more.

Marine Biology - the study of living things in the oceans and other watery environments.

Marine Science - a collection of sciences that focus on oceans and ocean processes at the water's surface, deep sea, coastal environments, and more including Marine Biology, Oceanography, and others.

Materials Engineering and Science - the science and engineering that focuses on studying materials and their characteristics at a structural level.

Mathematics - the study of numbers, operations, structure, space, and change; it's much more than arithmetic and geometry.

Mentoring - when a more experienced person helps guide a less experienced person in professional and personal journeys; mentors provide advice, resources, connections, and an understanding listener who can relate to similar challenges.

Mechanical Engineering - the engineering dealing with machines and how they work.

Metamorphosis - the life transformation process from an immature form to an adult form in several stages like a caterpillar to a butterfly.

Metastasis - the spread of cancer cells to other organs and systems throughout the body.

Microbe - a small organism made up of only one cell, often times a bacteria and also includes viruses and even some very small fungi.

Microbiome - the collection of all the types of microbes living in a certain host.

Molecule - a combination of at least two atoms bonded together to form a chemical compound, the smallest unit of a chemical reaction.

Nanoparticle - super small particles (less than 100 nanometers) that can have different properties than larger particles of the same materials, sometimes used in technological processes.

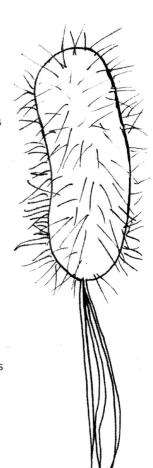

Nervous System - a complex collection of nerves and specialized cells, called neurons, that communicate signals with different parts of the body. The main organs involved in the nervous system are the brain and spinal cord.

Neurobiology - the study of the nervous system and the brain; sometimes called Neuroscience.

Neurosurgery - surgery that focuses on the nervous system like the brain and spinal cord.

Nuclear Fission - when the nucleus of an atom divides into smaller parts; can be caused by either a nuclear reaction or radioactive decay.

Oceanography - the study of the oceans; there are several types of oceanography including Chemical, Biological, Geological, and Physical

Organism - a living thing such as an animal, insect, plant, or microbe.

Optics - a branch of physics that focuses on the study of light, how light behaves, and sight.

Outreach - the act of bringing science to a broader audience to inform, excite, and inspire.

Patent - a special license from the government that allows the inventor of a product to have the exclusive rights to the material design and production of that product.

Pathogen - a disease-causing microbe.

Pathology - the study of the cause and effect of diseases.

Pediatric Neurosurgery - surgery that focuses on part of the nervous system like the brain or spinal cord, specifically performed on babies and young children.

Pharmaceutical - relating to medicinal drugs and how they are made and used.

Physics - the study of matter, specifically how it moves and behaves through time and space.

Physiology - the study of organisms' bodies and systems.

Plankton - organisms that live in the water that cannot swim against a current, they can be small animals, plants, or even larval versions of fish and other ocean animals.

Plastics Pollution - the build up of plastic products, or pieces of plastic products, that can negatively affect wildlife and the environment.

Polysaccharide - complex sugar molecules.

Prefrontal Cortex - the part of the brain that controls thinking, reasoning, impulsiveness, and more.

Primatology - the specific branch of Anthropology focusing on primates such as humans, monkeys, and lemurs.

Professional Development - activities for professional growth; in science professional development often takes place as workshops, symposiums, and conferences with peer researchers, science communicators, and more.

Protozoa - a single-celled organism with capabilities like mobility and predation.

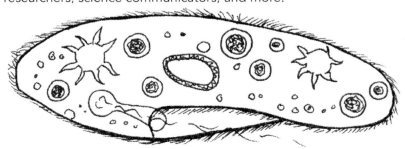

Protype - a sample version of a product that can be used to test the product's design, function, breaking points, and more to make improvements before going to production.

Psychiatry - the branch of medicine focused on the study of mental disorders and how to prevent or treat them.

Quality Assurance - in technology, this is a form of application testing.

Radiation - high energy waves that cause ionization which can be damaging to health.

Real Analysis - the theory of real numbers and real-valued functions.

Reproductive Physiology - a component of Animal Science that studies how animals have babies and looks into possible ways to develop technologies that help animals reproduce when having difficulties.

Rover - a vehicle designed to move across the surface of a planet in order to explore and often times collect data in photos, samples, etc.

Scale - the relative size of something; often talked about in science because STEM deals with super small things like atoms to super big things like galaxies.

Science Communication - the communication of science to a broad audience, usually non-academic and non-scientist, that focuses on understandable concepts relating to other people and groups.

Security Engineer - also known as a cybersecurity engineer - a computer engineer who focuses on protecting computers and the data they store or access.

Sediment - the collection of matter that settles to the bottom of a basin, like the collection of dirt, shell fragments, and more that settles to the bottom of the ocean floor.

Sexism - the prejudice and discrimination against a certain sex or gender, usually against women.

Software - the set of instructions that make computers and computer programs run.

Software Engineer - an engineer who designs and writes the instructions that computers use to operation.

Spatial Graph Theory - looks at how collections of points and edges can be positioned symmetrically in a three-dimensional space.

Spectroscopy - the study of how matter and radiation interact.

Star - a celestial object made up of a sphere of plasma that glows and is held together by its own gravity, the closest star to planet Earth is the Sun.

Statistics - in simplest terms, the study of data which includes the collection, analysis, and interpretation of data.

STEAM - the acronym used to note science, technology, engineering, arts, and mathematics, important for including the importance of creativity in STEM work. The STEAM movement works to include the arts as a valuable addition to STEM work.

STEM - the acronym used to note science, technology, engineering, and mathematics.

Structural Biochemistry - the study and application of how molecules are shaped and interact with each other.

Subsistence Lifestyle - a way of living where you collect and prepare your resources from the land in ways that respect nature and cultural traditions.

Supermarginal Gyrus - a super cool (couldn't help myself!) part of the brain that is important for reading, it helps the brain connect letters into words.

Sustainability - efforts and actions taken to support long-term balance in the environment and our planet's ecosystem, examples including reducing the use of natural resources, alternative energies, etc.

Systems Engineering - See Industrial Engineering.

Technology - the knowledge that focuses on how technical means are created and used in connection to people and the environment.

Telescope - an instrument that uses various curved mirrors and lenses to magnify distant objects.

Thermal Degradation - the heating up of a material that causes the internal proteins to break down into their smaller parts.

Topology - a field of mathematics focused on the properties of space.

Qualifying Exams - the major exams that students must pass to get into their graduate programs.

Quantitative Genetics - the specific type of genetics that studies how traits are inherited.

Quantum Dots - tiny crystals that have the ability to emit energy.

Undergraduate Studies or Undergraduate Work - the first stage of a college career, the first tier of academic degree.

Veterinarian - a doctor that helps animals or does research on animal systems.

Watershed Science - the study of natural processes and human activities that affect freshwater resources.

Zoology - the study of animals; either Vertebrate Zoology (animals with backbones) or Invertebrate Zoology (animals without backbones).

ACKNOWLEDGEMENTS

This *Super Cool Scientists* journey has been an amazing ride!

There were so many people involved in making the original *Super Cool Scientists* book a success. The Kickstarter supporters, the super cool featured scientists who inspired me with their enthusiasm and helped promote the project, and my personal cheerleaders.

That energy from the original book pushed me to create *Super Cool Scientists #2*.

Again, I have to thank illustrator Yvonne Page first. She has truly been the perfect partner in these projects. It has been a pleasure working with her and I'm honored to get to share her artwork with others as part of *Super Cool Scientists*.

Second, the incredible group of women who we feature in this book. Working with this second cohort of super cool scientists has been invigorating. I learned some new science concepts and had a lot of fun in the process. These women do such important work each day and continue to give back to others along the way.

Finally, I'd like to thank all my people. The time between publishing the original book and this one has been an incredibly challenging one for me both personally and professionally. I've made it back to a joyful place thanks to my crew of supportive friends and family. Some of them have been in my life since the beginning, some I've only just met. I needed all of them for this stage of my journey and I thank you deeply.

I truly believe that the secret to a joyful life is surrounding yourself with good people. Invest in people who invest in you.

ABOUT THE AUTHOR

Science stole Sara MacSorley's heart the first time she visited an aquarium. She studied Marine Biology at the University of Rhode Island thinking she would become a researcher. Instead, she learned that a science degree opens up many more career options and started on a path of science education and outreach. After earning a graduate degree in Business Administration, Sara moved into project management and leadership roles in higher education.

Now, she runs Super Cool Scientists, LLC to celebrate even more women in science, technology, engineering, and mathematics through storytelling. Her passion to make STEM more inclusive is a theme in all her professional and outreach work.

www.saramacsorley.com

ABOUT THE ILLUSTRATOR

Yvonne Page is a freelance writer and illustrator. Her work has been featured in magazines and publications and she is the author and illustrator of the picture book *Stinky Poo*. Her work can be seen on her website, Facebook, and Instagram. When not drawing or painting, Yvonne's favorite thing to do is spend time with her family, their two dogs, and cat.

www.yvonnepage.com